THE FLYING BUSHMAN

A TASTE AND LOOK OF THE REAL BUSH

By Greg Keynes

www.theflyingbushman.com

"I saw my first sunrise in 1902, at least that's what they told me" said Frank Shaughnessy referring to where he was born under a tree at Coolyun Springs on Byro Station. His father was a teamster from South Australia who brought his bullocks to Geraldton WA in 1894 to work in the gold industry. His mother was a local Yamatji woman from the Murchison region.
- Myself and Frank Shaughnessy around a bush camp fire at Middle Camp on the Wooramel River trip 11th August 1990.
I have planned to write more about Frank and his exploits in future additions of my books, he was well respected by both black and white Australians because of the example he showed, as a hard working non drinking, non smoking quality human being who had a very tough life, but never showed any animosity.

To my parents who gave their all for their family in a tough and rugged environment where Mother nature might kiss you or smash you and you don't always know which is which until much later.

Foreword

I have a passion for Agriculture and in particular the Pastoral Industry in the Murchison region of Western Australia where I grew up. It concerns me greatly to see the demise of the country where my family spent sixty years of sweat and effort in the production of many thousands of head of livestock and wool production over many positive years. Where the local communities thrived and employed large numbers of staff and contractors, rich with Australian character and endeavour.

Now there is little left to give your children the opportunity of a decent education in far away boarding schools to learn of the world and broaden their thinking. Thanks Mum and Dad.

As an individual now I can only stand idly by and watch the industry deterioration with apprehension, the lack of interest or care from Governments and watch the change in deteriorating weather conditions (be it climate change or not –it doesn't matter a damn) and the lack of financial prosperity drive good quality Australians away from what used to be magnificent productive lifestyle.

But collectively people can make positive change, and by way of highlighting the plight of this Industry and its challenging future, I have recently written a number of short bush stories below, along with some photos that I would like readers to consider, regarding what the bush was like, and perhaps enrol them in improving its future.

With awareness and understanding there comes the environment for positive change.

I sold Ballythunna Station my home for twelve years in 1997 – when there was still an asset to sell, but it doesn't stop me feeling for those still trying to raise a family and do their very best in the most trying of conditions, that are left behind.

Station principals who barely receive a mail service any more, and after purchasing their properties for significant funds are forced to work on the local Shire to put bread on the table.

What does that say to you??

I think we are inclined to forget –these are the type of people along with their forefathers that helped to build this country and helped make it what it is today.

I must say that what I have written in the following pages is completely my own personal opinion and not necessarily the opinion of other persons in the Pastoral Industry, and some people may not entirely agree with my comments. That's entirely their right in a democracy.

Another reason for producing this book is to show and explain how my home country, the Murchison Region of WA and surrounding areas to people who may not have had any exposure to this environment and enrol them to understand more about it and its challenging future.

For them to become interested, without any prior contact or firsthand experience, I believe they probably need the bush experiences related to them in such a way, from not only somebody who has been there, but somebody who can make the reading as interesting and relevant as possible to today, to enrol them into a better understanding of the bush.

I was fortunate to actually be involved in initiating helicopter mustering (as a pilot and Business Principal) into the southern Pastoral rangelands of the Pilbara, Gascoyne and Murchison Regions in WA since 1982 when I was 25, so I am well aware of progress in the pastoral/livestock industry.

From personally experiencing the "droving days" on a lesser level, to mustering huge areas of country single-handedly and yarding up thousands of cattle, wild goats, sheep etc – and pushing the gate shut with the chopper skid gear just because you could as per the photo below.

Myself yarding up at Mount Phillip cattle yards in 1984.

I also write about slightly different experiences that the bush gives us, like the strength of family, local community and particular bush characters.

Unfortunately most people won't have experienced the positives that this natural environment holds, but you know that's OK too – because we have moved on and things are different now –nothing stays the same.

I write for example, about my Mum as correspondence school teacher who taught her own four children for 22 years of correspondence schooling in the bush, without any mod cons, who used to require her students to address her as "Mrs Jones" in class, along with all the other imaginary school friends.

And thus my work is very much more relevant to a city based readership also, to enrol them into the bush experiences, so that they can better understand just what is/was involved out there in a very different environment, that most have not experienced. We all used to travel to "the farm" for the long weekend or Easter, but for all sorts of reasons – that doesn't happen that much anymore.

I write of another story, the only nonfiction story in this book that I made up in 10 minutes - my nightly story telling prior bed to my two grandchildren.

It's about a older sibling (seven years old) who wants to get back at her younger brother (four years), who hands him a tomato leaf from the back garden to use when the toilet paper runs out - the spines of the tomato leaf do their thing and he proceeds to chase her up the street in anger, much to the amusement of the morning traffic and local policeman

So they are a slightly different take on the old bush droving stories -but still have a very country flavour – and hopefully more relevancy in today's market.

My overall goal here is to not only become a successful freelance writer - but nearly as importantly, be able to sell the bush and its relevancy to the population at large - and in so doing, hopefully I can achieve more for Agricultural and remote Australia and educate the world population of its importance to our future.

We understand change, and aware that life moves on, but if we invest in education and understanding, technology, experience, finance and business interest back into this natural environment, I believe it can repay us tenfold. And we won't have to buy all our food from China in years to come –and our grandchildren will have a future in Agriculture in Australia and elsewhere.

If we don't make the effort and do something, somebody else will, and they could well be forgiven in saying, "well you had your chance and blew it". Why didn't you do something when you had the opportunity and where withal before it became too late?

Hence some of the reasons for this book.

I hope you enjoy.

Myself, the author, with my lovely Granddaughter Jasmine, who reignited my pleasure to write and recommence storytelling Christmas 2011.

Note from the Author

I just wanted to introduce myself to all you people across the world from me in Australia, especially to those of you who have a passion and background in Agriculture and the rural industry, along with its future.

I have just made the decision to place some of the above Australian nonfiction short stories onto Amazon.com, but more important than that, my goals and reasons for writing them, which I am sure many of you will certainly identify with. You will no doubt share my concerns regards the future for our children and grandchildren in Agriculture certainly in Western Australia and if not Australia, but very possibly, across the world.

I don't profess to be an expert on what's happening elsewhere in the world of Agriculture, but certainly my belief is, in Australia its terms of trade have diminished significantly, and the real producers are getting screwed to death, be they large or small.

Where once the average farmer could maintain a reasonable standard of living, and at least send his/her children away to a good education, that option now is extremely difficult.

I am using these stories as a launching pad for all primary producers across the world to unite together to avoid a continuation of the demise of Agriculture and its associated small businesses on the world stage, which is our most important resource I believe - as you can't eat rocks, and we can't keep on importing everything we eat from China.!!

This is on a backdrop of where in many cases our country is one of the most productive in the world, per unit of rainfall, and its producers are known worldwide for their production and skill, producing all sorts of quality produce, but we are not allowed to make a worthwhile living – what bullshit!!

There has to be a future in the regions of Australia and elsewhere, where people who are prepared to go out and bend their back and actually produce something for the plate and their children's future must be able to continue with financial security to employ progress and enjoy the fruits of their labour.

Whilst full credit to the people who are left, desperately trying to hold their asset together, hoping for better times ahead.

With all the challenges of potentially changing weather patterns, increased costs passed onto the producer, lack of government care or interest,(despite the window-dressing)poor commodity prices, dramatic increase in wild dog populations, increased bureaucracy, and animal liberation activists who act more on emotion than on any form of understanding or experience, life on the land has become financially intolerable. And the regions have become empty shells of their former selves.

I am not into this Tweeting and Face book and others, but if that's what we need to do to get our word out there then I'll give it a go. So please help me spread this declaration

Please share this amongst your friends and neighbours, not only to read these stories I have written so that you understand some of my background better, but that collectively we take on this challenge head on and draw worldwide attention to what needs to happen to bring Agriculture back into the black.

I urge you to take up this Banner of my writing name –the Flying Bushman, and let us all vote with our feet to make positive change for our children's future.

I will be available on the Flying Bushman website shortly as a site to express your interest and suggestions. Let's make 2014 and beyond a great time for a positive change in Agriculture and the regions of Australia.

Best wishes to you

Regards,

Greg Keynes (the Flying Bushman)

Maps of Australia

The Land Down Under

Australia: The state of Western Australia is on the far left.

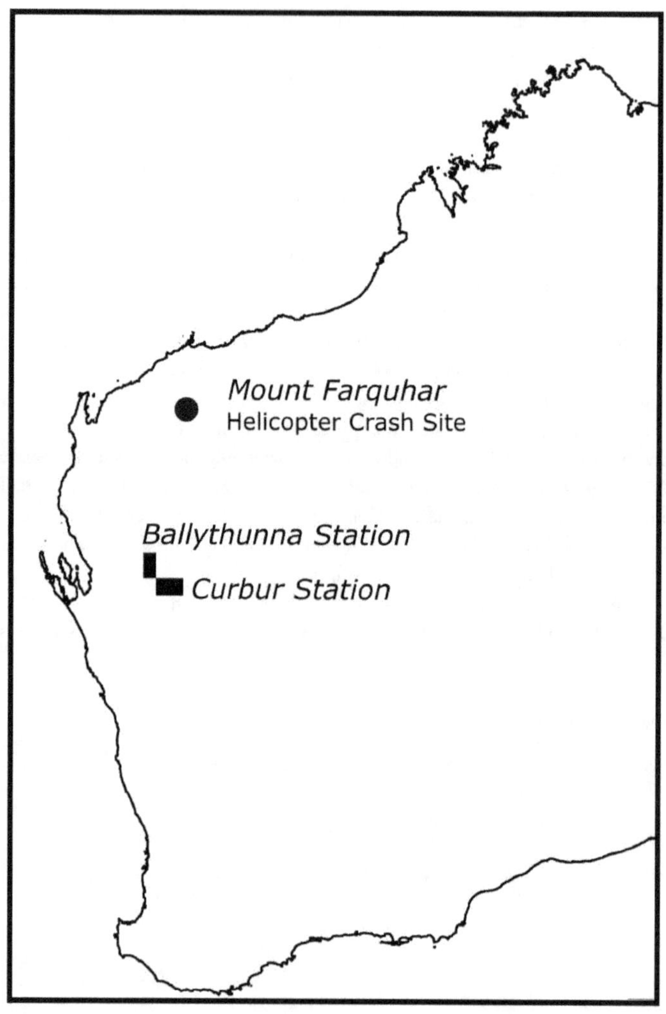

Closer view of the top half of Western Australia showing Curbur Station, Ballythunna Station and my crashsite of 1984. Note: locations are approximate

Prologue

We were still five hundred metres away, but the picture was looking grimmer the closer we got.
It's hard to explain in words but the visual site is awful. The emancipated sheep are walking very slowly, if at all in the heat and their normal running about is out of the question. Their flanks are hollow and their bare shorn skin seems to be drawn into their centre – like a magnet has sucked all the moisture from them and left just an empty shell of skin and bone. The hide is just hanging off their hip bones, like rags off a scarecrow. Most camp in the shade out of the draining sun, but on this occasion..........

Contents

Chapter 1

Tomato leaf on the bottom

Now Janet wanted to get back at her younger 4 year old brother Peanut Johnson who happened to be sitting on the toilet this day when he realized – there was no paper left!!!

Now 7 year old Janet had had enough of her brother's shenanigans recently, and was hell bent on bringing him back into line.

There he was screaming to her from the back toilet to bring him some paper –and well, she just thought it was her opportunity to give him a bit of his own medicine!!

She rushed out to the back garden where the large tomato bushes were growing and grabbed the closest and largest leaf she could find. Running back toward the toilet she leaned around the corner, to hand her brother the pitiful excuse for toilet paper. She certainly made sure she wasn't getting too close – for quite obvious reasons.

"Use this" she instructed, casually leaving her younger sibling in a complete quandary, as to what he was meant to do with what she had given him!!!

Reluctantly, and with no other option, the young boy leant forward and did what a young man has to do, in such cases!!

Weelll - the repercussions were immediate and severe.

The spines on the tomato leaf shot into his bottom, and removed him from his seat in very quick time.

It was then that it dawned on him at this tender young age, that his older sister had tricked him –and his anger rose as he moved from the toilet, in no uncertain terms, towards his sister's bedroom without even pulling up his jocks or trousers. In fact he was starkers, as he was on his way to the shower previously, when nature had interrupted him.

Now his sister, obviously assumed something may be coming, and was organising her retreat via the front door, as Peanut rounded the corner toward her.

He nearly "got her" before she opened the front door, but was in hot pursuit and much closer by the front gate. Onto Mark Street footpath they went, completely unaware of the morning traffic. Peanut Johnson with the tomato leaf still stuck firmly in position was in hot pursuit of his sister, and closing the gap.

Within 100 metres of the North West Coastal Highway such was Peanut Johnson's anger that he was minimising the distance on his sister, and even less aware of the slowing traffic amazed at what they were witnessing.

With the tomato leaf still well attached and within grasping distance of his sister's long blonde hair, his attention was instantly taken by the siren and flashing lights from the oncoming police car.

The local Sergeant jumped out and with a wry smile on the other side of his face said, "Righto what's going on here!!!!"

It was only then, that Peanut Johnson began to realise his awkward position.

Doing his very best to keep a straight face, especially with the fast growing car loads of raucous spectators, the sergeant suggested in his very strongest voice, "that both siblings get along home and report back to their mother", as he wisely guessed, that their lodgings would be close by.

Both children, turned quickly on their heel and returned to whence they had come, more aware now of the commotion they had caused, especially considering the "hooting" of passing car horns, and good hearted comments out the windows of passing traffic.

As Peanut Johnson was last seen entering the front door of his house, he was still muttering very angry comments to his sister; the tomato leaf was still vividly attached!!

These are the two culprits, Jasmine and Robert from the story
above ready to go camping on the back of the Ute.

Chapter 2

When I nearly perished as a kid

It was pretty hot by the time we drove into Georges wind mill and it was near midday. Close to a century in the old temperature language, and as often at that time of the day in the heat the black crows were circling high in the air, on the thermals – especially if they have a specific reason. The site was pretty awful as we drove in and I could hear Dad swearing low under his breath, as experience in the bush gives you an inkling of what you may well expect to see.

Dad always hated the situation of animals suffering, because of mans actions, directly or indirectly. He used to say, we put the livestock in this position and it is our responsibility to make sure they're looked after. It was something I always respected about him, although we didn't always agree on everything. But you don't really have a say do you when you're six or so. Even if you figure you may happen to be right. As we got closer, the sheep in the distance that came into view were very perished and dehydrated to say the least, and four paddocks ran into corners at George's mill, with fair numbers in each.

It's hard to explain in words but the visual site is awful. The emaciated sheep are walking very slowly, almost painfully in the searing heat and their normal running about is out of the question. Their flanks are hollow and their bare shorn skin seems to be drawn into their centre – like a magnet has sucked all the moisture from them and left just an empty shell of skin and bone, the hide just hanging off hip bones, like clothes off a scarecrow. Most camp in the shade out of the draining sun, but on this occasion some were hanging close to a pathetic little shade closer to the water trough, just made from a couple of lengths of inch and a half water pipe, running across the centre of the water trough to stop stock jumping from one paddock to the other, whilst drinking. They hung their heads close to the ground, and occasionally flicked their ears from the annoying flies, and that showed the only limited movement. We were still three hundred metres away, but the picture was looking grimmer the closer we got.

It would have been about 1962 on Curbur Station and it was a very hot summer. It was nearly Christmas and all the Aboriginal staff and stockmen had just got onto the mail truck for Christmas holidays that morning. I remember seeing them all cleaned up with the flash coloured silky shirts and RM Williams boots, that they loved and the smell of Californian poppy hair oil, ready to go into town and spend holidays with family in places like Mullewa and Geraldton – for the cooler summer sea breeze.

I had remembered the discussion Dad had had with Johnny the head stockman/overseer when he asked him about how much water was in Georges Tank, when he came back from the bore run only yesterday. "Yeah e orright", Johnnie had said smiling with the cigarette smoke whispering out over his thick black curly beard. "He pumpin right, but e not much wind intee". I was pretty small but I had helped pull and primed plenty of windmills with Dad and knew that if they lost their water, often they wouldn't start pumping again, unless there was a real good wind. "But big mob there too intee, and he flat out keepin up ah", Johnnie had said, as if the situation required extra explanation.

"But everything had got a drink though", Dad had enquired.

"Yeah, he right, he good" was Johnnie's reply.

That's why we were so late going out you see, because normally you check waters from first light, especially in the summertime when it's cooler. The mail truck had left with all the crew and families on board all excited and ready to party for Christmas around mid morning, but Dad was unsure as to the state of play at Georges in his gut, and the fact he was the last one there prior Johnnie's visit nearly a week ago. As the head stockman had said there was a big mob watering there. Dad obviously just didn't like it, and the fact that the mail truck was on its last trip before Christmas – just added a bit more angst in his mind. Nobody would have wanted to be too late for that last truck ride, by whatever means.

So we drove into Georges, from the ironstone rise to the west and on arrival surveyed the demoralising tragedy, some ten or fifteen sheep were already dead, mostly their heads stuck in troughs near ball taps to suck the last few drips. Some emus rushed away from the area and seemed like the only animals with any energy. Crows were around immobile carcases devouring and removing available goodies, but lifted in black clouds, to nearby trees on our arrival.

We guessed what the problem was well in advance. Dad's gut feeling had been substantiated and the words he was yelling are not up for print.

It's the feeling of helplessness, as you watch the misery of the damage that has already been done previously days before, with bleats from animals in pain and anguish, calling for moisture and not running away from us as they normally would. Only meters away in fact, wild pastoral sheep, that only see man once a year and I was nearly able to touch them, showing no concern other than their need for water and survival.

I pulled a few immobile carcases out of the trough by the back legs, while Dad was checking the windmill, hoping they were still alive, but they had gone to sheep heaven. The wool on their legs oozed away from the carcase into my hands and I was forced to get another grip on the bone instead of the outer skin.

Dad said, "Grab that waterbag out of the jeep, and we'll see if we can prime this bugger". As I returned from the vehicle, with waterbag in hand, two ewes came towards me smelling at the cool drinking water in the bag. Their reaction was instant and self preservation was they're only consideration. It was only then I realized the choir of "baaarring" noise they were all making, perhaps from the smell of the water, in seeing the two or three ewes run to me, or perhaps the chance of us been able to provide a miracle for them. They had experienced man watering and feeding them in times gone by, on desperate occasions, although perhaps not as bad as this. They're not as stupid as some people think, especially when it comes to their own survival.

The windmill was hardly turning in the feeble breeze, as Dad poured the water from the bag into the top of the column, but then it stopped completely as the load came on the head.

"Just give it a prime for me", he said, automatically expecting me to climb up the ten foot ladder and hang over a sixty foot well, whilst pouring the contents of the bag into the top of the column. He went up above me another ten foot and climbed out onto the windmill platform (just a little narrow wooded frame to walk on whilst repairing the main head). He was standing alongside the spinning eight foot diameter Southern Cross windmill head, increasing its revs by hand, and in turn causing the rods in the column to lift up and down quicker – hopefully producing water as I poured water into the column to prime it.

He had already banged the column to loosen a clack in the pump deep below if needed, so now all as we could do was hope. He pulled and pulled on the metal spokes supporting the head spinning it faster and faster, but after lots of swearing and sweat dripping from the tower above me, we all realized our worst fears were materialising.

And the sheep, realizing the new potential opportunity, kept baaaaaaa-ing in harmony and hope in support from the ground below. There must have been around eight hundred head in the four paddocks and the noise was becoming quite deafening. They were hoping for and expecting results and pushed closer around us and the mill, to get close to the action and more to the point be closer to the trough –when hopefully the water came through. Perhaps in a subconscious way believing Dad's noise and swearing, the speeding of the windmill activation and the general increase in energy around the site –should have achieved results – some trotted with difficulty and ungainliness towards the trough and the disappointment showed in the resultant body language, on their arrival and their close inspection confirming no moisture.

But you can't get blood out of a stone.

It was no good; we weren't going to get a positive result that way. Disappointment and frustration epitomised Dad as he slowly climbed down the windmill tower. "Bloody bastard", he said as his feet hit the ground, "we'll just have to move em".

Little did I know then the ramifications of that decision, and the following experience that would be with me for the rest of my life!!

Well of course I knew what "move em" meant, we would mob them up and take them to the next closest water, maybe some 10 kilometres away down the fence – but as I was to find out later – talking about it was the easy bit!!

We each had a good belly full of water out of the water bag I was still holding, and surveyed the scene for ways to accomplish this bloody awful job ahead. Sheep were still making lots of noise despite our lack of accomplishment for them, as if to guess we had a plan B

"We better take em to Tin Hut," Dad said, "Inneywobbing is too far, and there's too much scrub that way". I was a little disappointed because I had always liked the old Inneywobbing out camp, just an old corrugated iron hut where fencing and dogging contractors used to camp whilst on the job. I remember when the Yamatjis were camped out there for shearing mustering with horses, they had a little veggie garden going on the limestone, with a little low fence surrounding the house. They used the water tank overflow pipe for watering the watermelons and rockmelons they had growing there, so I was a bit disappointed we weren't going that way.

But I understood the bit about the thick scrub along the way – and we only had the jeep, and "Lady" the old Kelpie bitch was getting older.

It was my job to open up all the four gates to mob the sheep up in one mob to head north to tin hut around five or six mile I guessed –I'd been that way with Dad quite a few times on the mill run but whatever, we had no choice that was it, otherwise they died which was not an option.

Now many farmers had water trucks, or big tankers and pumps in the agricultural areas, but on the pastoral stations in those days nobody had much of that type equipment. With the regular checking and maintenance this situation should not have occurred – but shit happens, we just had to fix it now.

Dad and Lady had the sheep from the south western paddock "Moolarie", mobbed up and gently moving through the gate into "Curbur" paddock. There was not the usual rushing, not only because they were all so very dehydrated, but because it was bloody hot around forty plus degrees centigrade and sheep just hobbled together into a mob reluctantly. Of course there was a few sheep that were quite sprightly, obviously been off water for much less time, perhaps they just came in for their first drink this morning to find it dry. They were to be very responsible for the later dramas to come, but they were only looking after number one, and could not be blamed.

We finally got the combined mob out of Curbur paddock into Tin hut paddock (that's the paddock not the mill –the mill is on the NW end of the paddock) and the others out of Muthernabbin and after twenty minutes or so we were mobbed up and ready to head north towards Tin Hut mill. Travelling on the eastern side of the fence line where the mill track was, some ten kilometres away to the north.

It looked OK to start with, as the animals were at their strongest initially that they were going to be, and of course we weren't expecting speed and knew well it was going to be a hard slog. Importantly going north from Georges towards Tin Hut its quite open country for a kilometre or so, and everything was going along pretty well, or so we thought.

However after and hour or so the old bitch Lady was knocking up understandably, and the tail was constantly pulling up in the shade now that we were starting to get into thicker shade country. While this was happening, the stronger faster young sheep I mentioned before were easing out away from the fence to the north east, for protection into the bush, attempting to slip away from Dad in the jeep.

You see they didn't have the same need as the others for water at Georges, but of course were still thirsty, and to avoid a break in the wool fleece they ideally required water also, but their endeavour for freedom was stronger than their need for water at that particular point in time.

That's about when Dad made the near fatal, but understandable decision to ask me to " keep up the tail", as the tail animals were pulling up under the attractive shade trees, and getting further and further behind, and I jumped out of the jeep to push up the shade hugging stragglers. The mob was affectively splitting apart in the scrubby country where you could only see fifty to eighty metres in front of you, and vehicle access and manoeuvrability was becoming more difficult.

Neither of us thought this was a problem, me getting on the tail of the mob as I had done this before and the fence was just over there as a guide anyway, as Dad moved the vehicle up and down the wing to keep the wanderers in, and through the scrub I could hear him continuously, so there was no problem initially.

A few things happened though that we didn't expect.

I was initially trotting on the tail of the mob, which was probably stretched a hundred meters or so across and in sight of the fence through the trees. Initially the sheep moved well from my close proximity and perceived pressure, but after a few kilometres they took less notice and were pulling up more in the shade, and becoming less responsive to my pressure. At the same time Dad's jeep noise was getting further away as his sheep on the top wing (those fit bastards) were pushing out further towards the NE away from the fence –and me.

I realised I was keen for a drink from that waterbag in the jeep after an hour or so of pushing the tail –although now I had sensibly eased down to a steady walk. One might say the excitement of doing a big man's job was wearing off!!

Perhaps another fifteen or twenty minutes went by since I had seen the jeep in close proximity and I was getting desperate for Dad to come back to the tail of the mob. But he didn't' as he was obviously occupied trying to keep the top wing in some distance away

The other thing that was happening slowly that I didn't appreciate until after, was that the tail mob I was pushing was getting a further and further spread from the fence. It got to the stage where the sheep on the outside tail were way off the fence and hanging back in the shade –I nearly missed some because they were out of sight, and I had to run back and over to them. I didn't know how far I was from the fence at that stage, and after giving these stragglers a stirring push in the right direction –headed back towards where I thought the fence was.

It was becoming very hot and steamy by then, and whilst we had been involved in our job the clouds had quickly built up which I had not noticed and were by now pretty much blanketing the sun, making direction in the scrub harder to confirm. Large beads of sweat were standing out on my arms and my shirt was drenched from the running, and my mouth was now becoming dry and thick against my tongue, walking from shade tree to shade tree, taking a leaf out of the sheep's' book.

I kept on heading back for the fence through the mulga wash, pushing sheep along as I was going, but after travelling for what seemed a considerable distance and time –I still hadn't hit the fence. Why not? Was I going the right way? I was sure the fence was that way!

I couldn't hear dad's jeep at all now, and hadn't done for some time, so I was rapidly losing confidence in my direction capabilities!

This was the scariest experience, especially as a young boy, and that is when you are trying to come to terms with the fact, that you may be travelling in a very different direction (perhaps the opposite) than what you originally thought. You feel like the natural compass inside your head, which always points north has been lost, and then the terrible understating creeps on you that the direction you thought was north – is not north at all –in fact you are not sure what direction is what.

I couldn't hear the jeep at all now, and hadn't done so for some time! Had he stopped or broken down or was he that far away I couldn't hear him and what direction should I be listening for him anyway?

It was then that I realised just how thirsty I was and for the first time, lost some interest in getting the sheep to water and was starting to think more and more of my own survival.

I sat down in the shadiest spot I could find handy, and tried to come to terms with the awful event that had transpired, trying to think calmly although my mouth was so dry, and my tongue was feeling like a large dry rock in a cave. I strained to hear the noise of the jeep –in any direction. There was none and only silence came back from this bush oven, only the noise of the odd crow could be heard and I shivered as I thought of what they had done to those sheep back at George's mill. Even the birdlife had gone to shade cover and not flying to save energy in the shimmering heat, the tail sheep had stopped making any noise as they were now also stationary in the shade in close proximity.

It was now I was starting to become desperate and losing my cool, I knew things were fast becoming really dire, as Dad was a long way away and most of all I had lost confidence in the direction I thought I was going.

The desperate concern was overtaking me now and I was starting to feel sure that I may not be found in time, as how and why was Dad going to find me in this thick scrub where I was presently, as I had nothing to guide him to me? No matches, no smoke from a fire. I had jumped out of the jeep without considering that in my youthful enthusiasm.

I broke down and started to cry, in the inadequate shade of an old mulga tree. When it gets that hot, there never seems to be a decent shade anywhere in the bush, just modest reprieve from that blistering sun. The tears filled my eyes uncontrollably and I felt so abandoned. I wasn't sure how there was enough moisture in my body to actually cry tears but I did, and I gratefully licked them off my dry hands to gain some very limited moisture onto my parched tongue which evaporated into the desert of my mouth.

Practically, through tears now, I was trying to weigh up my options which were scaring me to think about. Nothing too rational and sensible seemed to be coming through as my nose was starting to run and tears were still pouring down my face as I was trying to come to this awful bloody realisation –that I could just die right here –and they may not even find me in this scrub before the crows cleaned me up.

I flopped to the ground and just wilted into a teary mess, desperately required moisture dripping onto the hard Mulga wash ground.

I do remember screaming out a few times for Mum (who was back 30 kilometres away at the homestead completely unaware of what was happening – she would have been completely beside herself with worry) and really just completely giving myself up. The desperate crying and convulsions had shattered the core of my conscious being it seemed, or outer exterior of my being at any rate. In that moment I gave myself up to the universe, or God or whoever was in charge of this show.

A little while afterwards and as surprising as it sounds, I felt slightly relieved after crying myself out, and more light headed about the dire position I knew I was in. And perhaps I had actually come to terms with the fact there was nothing else I could do. A pretty hard lesson I guess and a peaceful state seemed to engulf me, which may be described as an out of body experience, as if I was looking at myself from outside this awful situation, as if somebody or something was helping me to come to terms with my position and trying to give me some positive perspective and reinforcement.

I hadn't been able to hear the jeep for ages it seemed, and because I was so exhausted and not exactly clear on what way to go anyway, I'm sure I just walked over a little distance to a better shade tree, and laid down to rest and contemplate my situation. But there wasn't much to contemplate really, just in your face bloody awful reflection this wasn't somebody else in this movie – this was just me, right here right now.

I didn't even know how long ago I had my last drink at Georges with Dad, and some tears slipped out as I realised it may be the last drink I would have with him –then I stopped myself and said no it couldn't be, and then reverted back to my previous light headed space of calm.

I can't be sure, but perhaps a little of me died that day, certainly perhaps a fair amount of childhood naivety anyway.

After some time, I'm not sure how long without any watch I decided to get up and walk back towards where I believed the fence to be, in this slightly improved state of mind, because I gathered that nobody else would help me if I didn't help myself – if I could get on that fence and mill run track that I would recognise (I hoped in my state) which way it was towards Tin Hut mill and water. The clouds were building and much darker now and that was at least providing some relief from the burning sunlight. To this day I'm not sure what gave me the strength or drive to lift myself up and again make an effort –an effort to survive I guess.

I was surprised at myself that I had lost interest in the sheep's' well being – as I knew they were so desperate for water too, as I now well and truly understood as we always looked after the animals first.

But I resigned myself that if I died it would be very serious – but if they died nobody would know. I also thought of the pain my Mum and Dad would have –Dad putting me in this position and Mum not knowing and finding out the hard way. I think that gave me some added step in my gentle walk toward what I thought should be the fence line direction.

I don't know what distance I travelled or how long it would have been, but eventually, although I hadn't reached the fence, I could hear the faint noise of the jeep – a long way in the distance. There was a desperate urge to run towards the noise, but common sense grabbed me and I started walking more slowly towards the vehicle noise, but with considerably more spring in my step.

He was certainly coming closer, not directly but the noise was getting slightly stronger and I could hear he was heading roughly in my direction. I started to quicken my pace, and although terribly thirsty and dry, the clean mulga stick that I had selected to chew on was now giving my burning mouth and tongue some slight relief it seemed. It was so dry prior the stick, it felt like my tongue would get caught on the inner side of my cheeks and the floor of my mouth, making swallowing nearly impossible from the dryness. The sucking on the stick was one of the positive ideas from my previous meditative state that I slipped into, when I collapsed under the tree.

I had walked maybe half a mile towards the noise in the opposite direction from what I thought was the fence when I came to a complete standstill. The noise was fading and was certainly getting further away.

Jesus I didn't believe it. Come on God I thought, "You were meant to be looking after me! Back there I gave myself to you." I turned my head so one ear was in the direction of the vehicle to get better sound. I didn't want to comprehend or accept it, but the noise was going away, it was getting fainter.

I had walked toward what I thought was the fence before and had been interrupted by the jeep noise and turned around to follow that in the other direction. Thus not confirming in my mind if I was going in the right direction toward the fence or not, because I had not been able to successfully prove or establish its presence. So now I didn't have a bloody clue which direction was which? I found another shady tree and after sitting down, actually laid down and stretched out.

I was laying there for only God knows how long but with later details it must have been quite a while. Because it was much later in the afternoon when I was awakened out of a sleep or some sought of slumber, or meditative state to quite clearly hear the distant noise of the jeep. At this stage it was closer than it had been at any stage since we had got separated, and I felt a pang of excitement surge into my stomach. I sat up and surprisingly after the relaxation and peaceful rest, my thirst didn't seem as bad and I felt quite stronger and brighter, especially as the jeep noise was so much closer.

I think I must explain here that bush kids in the 60's were pretty tough little people, who were not only used to the heat and tough conditions, but were well used to completing a number of set tasks and jobs that were expected to be done. One was in pretty good condition I guess you could say.

So I even started to run, but quickly caught myself and settled back to a brisk walk. I realised it was much later and the sun had lost its brightness and heat and was heading down in the western sky – the western sky that must be a more westerly direction behind me and I was walking NE toward the noise –I think!!. I nearly had an overconfident thought that if I missed the noise and jeep this time, if I kept heading toward that sun I must hit the western fence. It was still covered by clouds, but the glow from the sun was coming from over there in the west. Then I lost my confidence on that idea and just hoped like hell I caught up with the jeep this time. This time it was do or die.

I remember thinking, if I don't manage to get Dad's attention in the jeep this time, I don't think my heart or soul could take it anymore.

He got really close, maybe only a few hundred meters, and still out of sight, but he was overshooting past me I could tell from the noise, and going back south west towards the fence. He was beeping the horn on the old Willy's and was yelling out my name as he was going through the bush, as I could hear the smashing of the bush under the vehicle and tyres. I was running now, pretty fast little runner I was, adrenaline was no doubt spurting into my veins. Then I could actually see the vehicle and Dad driving along, looking forward and yakiing for me, always looking forward but not looking back. Now I was directly behind the jeep, running as fast as I could as he was driving directly away from me. He wasn't driving that fast and I was running as fast as I could around the bush and over anything I could logs and broken branches, but I was not making headway on the vehicle. After some considerable distance he was getting further away and my screaming between breaths was not getting his attention. He was still looking forward and to the side, calling my name. God if I had a gun!!

He was slipping further away from me and I was becoming totally exhausted as my running despite my vigour was slowing dramatically.

Then it happened unexpectedly, Dad stopped the jeep and through my tears I could hear his yells without the background of the vehicle noise over it. Was I dreaming, the vehicle was out of site by this stage, probably a few hundred meters or more through the bush at least, but there was certainly a hush from the vehicle noise, and then Dad yelled again.

You don't reckon I yelled back to him, I roared like a bull I reckon, and started running again barely able to keep upright. Can't imagine in my wildest dreams what sort of sound I made, but I knew this was my last chance and I had to make it count.

I was still running as Dad and the jeep soon came into sight, as he was now stationary and FINALLY LOOKING in the right direction – BACKWARDS and he could now hear and see me.

My God, how good was that!!!

He jumped out of the driver's side and grabbed the waterbag as he did so, quickly walking towards me. I don't think we even hugged you know. Men are funny aren't they, but it was very different back in those days. I remember a relieved smile on his face as he passed me the waterbag and I drank greedily, with water overflowing the sides of my mouth and onto my chest. As I did so I think he mentioned something about "gee it's good to finally find you", as he had been looking for hours.

I felt like saying, "tell me about it" but my utter relief overcame any smart comments and it wasn't the time, and I continued to drink eagerly.

The water wasn't that cool actually in the waterbag, and it was then that Dad told me that he was forced to reluctantly go back thirty kilometres to the homestead to refuel as if he had run out; he would have not been able to continue the search for me.

Hell, that was why there was silence from the jeep for so long. It was travelling to and from the homestead many kilometres away. I think I would have died if I knew he was that far away from me. The only thing that kept me going was the thought that he may appear around the next bush.

We caught up on the last few difficult hours together before we climbed into the jeep. It was so nice to feel the vibration of that motor under me and know I didn't have to do anymore exercise for the moment and just ease back into the old foam seat for the journey home.

As it was so much cooler now and the sun was mostly on its way down in the western sky, having lost its heat, and there was a rain storm brewing further north past the Tin Hut mill where we were originally heading. On the way out we came across mobs of sheep in long lines heading north, on the smell of rain, as pastoral sheep do.

They would either pick up some puddles closer to the Byro boundary or drop by the water trough at Tin Hut on the way through. They can smell a rain shower twenty kilometres away. The kangaroos too, were on the hop toward nature's milk bar. As we came onto the fence line that we were originally meant to travel, lines of sheep were steadily padding north toward the smell of pending rain water, drawing each other along, out of the bush like little guides to each other.

As I watched them with complete empathy and understanding of the desperate thirst they were concealing, I reached for the waterbag and drank down large mouthfuls of water –just because I could.

That was a big day at the office, and when the old Jeep hit the main road, I eased back into the comfort of the seat to think of what could have been. We didn't talk much on the journey home.

I reckon I must have gone to the other side and back that day, through that shimmery wall as a little "fella", and twenty years later when I had a serious helicopter accident in the Pilbara where I was flying, falling 150 feet with my machine, I thought to myself – I reckoned I have been here before in this particular space.

There was a few moments of time, I'm not sure for how long, that I eased through that shimmery wall again back into that peaceful place, immediately after the accident – just prior my pilot Rodney Pulford from my other machine, seeing me go down came over to pull me out of a wreck of a Robinson R22 helicopter that was not worth picking up.

But that's another story.

I can tell you this, it's not a place you would yearn for, it's not a place you would deliberately go, but it's a place of enduring peace!! Total peace.

I'm not sure to this day if Dad told Mum the complete story about that experience, perhaps he may have left a few little details out, but I guess it's one of those stories where you can say, "all's well that ends well".

Well, sort of.

Chopper mustering lovely fat merino ewes and lambs on the Curbur Lake.

Keros my brother on the bike with his dog "Butch" steadying the lead.

From the cockpit: a strong mustering combination, chopper, bike and dog together.

The old abandoned Moolarey yards on Curbur Station, built early 1900's by hand with axe, ads and crowbar, now just protecting the mulla mulla in a good season.

At the Curbur shed: merino ewes in the culling race being drafted off shears. An important job to maintain overall flock wool and frame quality.

Chapter 3

Sheepers jeepers

It was late in the afternoon and I was with Dad driving around the mill run in the old "Willy's Jeep". The soft canvas sides of the old green jeep flapping slightly in the breeze as we drove along. Lady, the old sheep dog was hanging much of her body out the side, to get a good breeze and any exciting scents ahead, dodging the regular bushes flashing past, doing their best to poke their head in to the cab on the way through. Dad wasn't saying much in his usual fashion, as driving was pretty much a full time job on those old mill run tracks, dodging stumps and sticks to avoid punctures.

It had been a long day for a little bloke like me at five, as we had been out since first light and done the eastern side of the Curbur mill run, some call them a "bore run", which was at least a hundred kilometres and most of that was very slow because of creek crossings, gates, cleaning troughs and checking watering points for stock water. We had collected two woolly wethers tied up in the small back section of the jeep that we had picked up on the run earlier during the day, which had obviously escaped that year's shearing muster, and they would be dropped in the straggler paddock further down the road on the way home. There wasn't much room in the back of the Willy's soft-top, just enough for a dog, tool box and a few extras like an axe, fencing pliers and trough brush, and on this particular occasion the two woolly sheep took up any excess room available – hence Lady's far flung stance.

Not far from Tin Hut mill we swung onto the "main road", as we called it, because it was the main gravel road between Mullewa and Gascoyne Junction, and my old home Curbur Station just happened to be pretty much exactly halfway between the two outback towns. Well you could hardly call "The Junction" as we locals called it a town, because it was really only a few old corrugated iron sheds, in the 60's and the junction pub, nestled just off the banks of the huge Gascoyne River east of Carnarvon, with its gigantic white river gums along its banks.

I remember thinking, as we hit the new surface of the main road, how much more comfortable the ride was, after the endless twists and turns all day, (no seat belt then of course,) that Dad was required to do. Although it was quite corrugated in the centre it was so smooth and straight, especially if you drifted over near its edge, where the main traffic didn't travel. It also made it a lot easier for Lady – the constant gripping pressure was off. She could just smile her face into the cool breeze and enjoy the ride.

It was easy for me between gates and around water points, as I could just drift off in my own little world, and watch the bush going by. I could see kangaroos lazing in the shade of large Mulga or Bowgada (pronounced "boggada") trees as we meandered by, startled emus caught unaware by our passing took off flat out into directions unknown –even to themselves. I had seen them on many occasions smash into the fence line ahead of the vehicle, time and time again, just smashing the fence and themselves, but achieving very little. It got lady excited though, she enjoyed the shows. Pretty small head – not much room for brains I reckon! Birds not in close proximity were mostly unaffected by our movement through their environment, sung and squawked their musical tunes as we went by little dust eddies in our wake.

The afternoon sun though let me know what direction we were travelling as we hit the main road near Tin Hut mill, as it was streaming into the old drop down dirty windscreen and Dad reached around to the front, with his old hat in an attempt to wipe the worst of the muck off from the outside. There were no doors on the old "Willys," so entry and exit was real easy.

You could look out and the rushing ground was right there just below you, but I didn't ever fall out though fortunately.

We'd done a few miles I guess, heading south west over the tin hut creek, that runs into the main Brille Brille Lake system (Google earth this lake 270 kilometres north of Mullewa, just west of the main road- see pictures below). "Fifteen square miles of water when it's full", Dad used to say to the sceptical visitors/ tourists when he was expelling the virtues of the Station.

Dad and I spotted it at a same time, but neither made comment, as we tried to work out what this sight was in the far distance. "Looks like some tourists" Dad said eventually as we surveyed the oncoming, but still distant traffic. Now we all knew Dad was a "practical joker", but at the end of a long day for a kid – I had no clues of what he had in mind.

He startled me when he said "quick, quick, jump out", as he stopped the vehicle to an impromptu halt. I was wondering what the hell he was doing. Now I had seen him move pretty quickly at times as he was a younger man back then. But this was amazing, perhaps we had a flat tyre at the back on his side, and he wanted to get the jack under before it went down. Naaa it couldn't be that - he wasn't swearing. But the big grin on his face gave something away –but I didn't know what it was.

"Quick" he said again, bringing me back to earth, and proceeded by his movements and showing his intentions, to drag one of the huge woolly wethers out of the rear of his side of the jeep. Maybe he was going to untie the sheep and let it go – but this wasn't the straggler paddock, and why the hurry and that mischievous grin that I rarely witnessed. These sheep were huge to lift, even for a strong man, with ten pounds of wool alone and a huge five year old frame, of a pastoral merino. I noted Dad had got the wether out from his side of the jeep and was heaving to lift it into the driver's seat, just behind the steering wheel. "Come here quick" he said, as he motioned for me to come around to his side of the vehicle, to the driver's door and hold the sheep upright in the seat that he had been occupying.

I did as he motioned, and by this time although still not really understanding his aims, grasped at least, his perceived urgency. He did the same with the sheep on my passenger side of the vehicle, the huge animal amazed and with astonishment on its face, at its quick change of environment sitting upright in the passenger's seat, from its previous laid back relaxed posture in the rear of the vehicle, gave a big belch from the base of its belly.

It was only when we were crouched low, in the foot space of the jeep, and out of sight, that Dad put the jeep into second gear and with forward motion started us driving off down the road towards the oncoming tourists with the caravan - that's when the penny dropped!!

Looking backwards and up at the sheep, from the floor as it were, with legs tied, perched on their backsides quite comfortably in both driver's and passenger seats, they seemed as if they were in on the joke too. Both sheep were smiling and sitting back bolt upright, as you could have expected them to attempt to struggle, or at least kick in the new position. But it seemed to me they knew something was going down –and they were just going to enjoy themselves anyway. We had to hold them of course –just in case –but out of site to the oncoming vehicle

When we eventually got close and then passed the car load of incredulous tourists a few moments later, their facial expressions were priceless. It changed from been friendly tourist waving madly to the locals, with two board city kids in the back seat, on a lonely outback track. To an astonished "head flip", open mouthed at what they felt they had seen, as if to double check what they thought they had seen, first time.

Two sheep driving a jeep!!!

Now I can only imagine the conversation in that tourist vehicle for the next couple of moments, no doubt discussing what they thought they had just witnessed, but it obviously required a more detailed examination, as in the review mirror Dad could see them turning around just behind us, desperately seeking a reasonable explanation to what they had just seen.

By this time, Dad had let go of any of his intrepidness, and was laughing loudly as we turned the old 4WD off the road and into the surrounding bush and off into the distance.

As they could not pursue us with their van in toe off the road they were last seen, out of the vehicle trying to extend their necks, to get a better look at the disappearing jeep into the mulga, in a desperate attempt to confirm the jeeps occupants.

Now as funny as this story may seem in hindsight, I figure the funniest part of it could be, if you were a fly on the wall when the occupants were attempting to explain the experience to persons who were not involved in the event.

Can't you imagine this, years later, when the children visit their grandpa in the nursing home!

"Grandpa, did you say you saw, - two sheep - driving a jeep, along a road"?? Are you completely sure Grandpa?

Had you been drinking enough water??

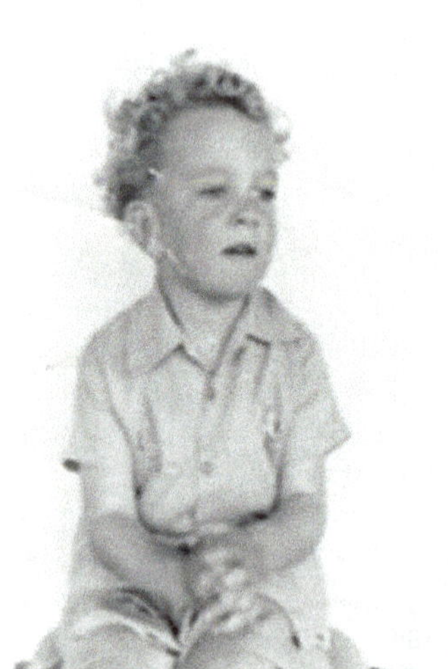

LEFT: Yours truly, Greg Keynes as a little bloke, who wasn't too much older than this when I nearly perished.
RIGHT: The Bush character, Les Keynes, my Dad who I think would have been a "natural joker on the stage" in another life. Unfortunately not many people witnessed this side of his nature.

Dad on the Curbur or "Brille Brille" lake in the dingy during a Sunday's picnic in the 90's.

A mob of sheep from the air at 1000 feet.

Dad at a family lunch gathering at my home at Wungong
WA after I sold Ballythunna Station in 1997

LEFT: The abandoned Muggon Shed shearing board, purchased by C.A.L.M.(Dept of Con. & Land Management) and neighbours of Curbur Station, my old home. Thousands of Merino sheep were shorn in this shed over the years along with many other Station sheds in the Murchison District. Now because of wild dogs and years of poor seasons and wool prices's, now there's hardly any merinos left in the Murchison at all.
RIGHT: The same Muggon Shed from the outside. Feel the abandonment and lack of business energy.

"Welcome Home" The cooks quarters with the kerosene fridge.

Wild feral goats that roamed the pastoral rangelands and were another form of positive cash flow. But now wild dogs have chased and mostly eaten them out of the country.

This is how goats are transported, many hundreds of kilometres away to local and export markets.

"Just relaxing in the shade", these little "kiddie goats" or capretto are offspring from the goats above, which are taken here to a Chapman Valley feedlot near Geraldton and cared for and fattened on the hay below and milk kibbles.

Haystacks to feed animals in the feedlot.

Chapter 4

Tony sitting on the back fence

Now Tony was the sort of guy who knew very well what he was meant to be doing, and how to do it, but only one time in ten would he apply himself.

Do you know of people like that?

But gee when he did, didn't he have the skill to blow your hat off. It was as if he used to save all the juice up for one big effort that would make you continue to employ him though all the very average efforts, but when he brought the team home, into the home straight, it was pretty to watch.

This is just such a story, about a very relaxed character, a very aggravated scrub bull, and a very special day in the cattle yards I will remember as a young man for the rest of my life.

We were all at Wheendong cattle yards in the early seventies, on the western side of Curbur about half a dozen of us, mostly Aboriginal stockman Dad and myself, when we ran this huge, wild, feral shorthorn bull into the receiving yard. He was rich dark red, and mean as hell (see picture below) weighing 700kgs plus probably, with forward pointing horns longer than normal, and he was constantly eyeing off everybody in turn as if to say, "I'm watching you, your turn will come". Experienced stockmen are loathe to state the obvious on such occasions, but one of the boys said in a low respectful tone, "watch dat bulla" (which is Aboriginal slang for watch that fellow) to all within earshot. He didn't really have to say it, but he did in a nervous respectful sort of way!!

The thing was we had done the easy bit pushing the mob into the receiving yard, now we had to get him into the forcing yard, where in turn, they run up the raceway and into the head bail or crush to be earmarked and tail tagged or whatever, for trucking to market. Getting him and the rest of the mob from the 40 by 15 meter oblong receiving yard into a much smaller triangle shaped forcing yard, a tenth of the size, was the challenge. Not the physical challenge of doing the job –that was no problem, it was doing it with your life still intact at the end of it was the challenge. Dad as the Boss, put the emphasis on the yard building materials been sourced as cheap and handy as possible and although usually strong, were not always the most user friendly, when it came to requiring quick exit out, through, or over yards. Gee those depression years had an impact didn't they??

The yards were rock solid made of two inch pipe running horizontally around the top, supporting heavy duty mesh with railway iron uprights on the outer side of the mesh and rails.

It was fortunate the yards were built strongly like this in one way though, as soon as we had yarded this mob with this particular bull, he realized the net was closing in. With panic emblazoned in his big wide eyes, he trotted indignantly around the new barriers to his freedom. One can only try to imagine what it would be like for a three four, or five year old "mickey" feral bull, which had grazed the rangeland all his life to be in this strange, dangerous restricted yard for the first time in his life with all these little scary humans in his space. He's not a stud bull but one that has escaped muster for some time, very probably for good reason. Coming to terms with that and attempting to deal with it, in an animal's mind must be a like a bush kid going to boarding school for the first time.

One of the boys opened up the forcing yard gate for access and we skirted around the rear of the mob, everybody very, very conscious of where "the bull" was. There was some other smaller mickies in there too, in this mob of 120 or so head of cattle, which at the moment weren't causing any problem, but they could easily become "fizzie" to if pressured too much.

Steadiness is the key as every cattleman knows, however when a bull like this bloke "wants a go" there's not very much one can do. Even if you stay way back on the back fence and don't move, they'll come at you for no reason when they're hot. Just like a drunken raged fool who just wants to fight. "Crusin for a brusin", one might say – either you or he may get the "brusin"

So there we were, the gates open and although we understand how this bull maybe feeling, we had a cattle truck to load and a dollar to make.

The swirling mob is "towie" and frightened and we apply only slight pressure or advancement toward them from the rear of the yard. The truth be known, nobody is wishing to advance too far from their respective closest available fence refuge, in case this big boy makes his charge. We all know it's only a matter of time, just a case of which one of us will be lucky!!

A couple of quiet cows do us a great favour and move into the forcing yard to offer our large friend an option to follow, but he refuses and breaks back dropping his head and blowing snot from his nose and a huge "burrrr" at one of the boys as he tears past, who quickly took advantage of the fence. As the rest of us rushed forward to take benefit of the lead and shut the gate on our first bite into the forcing yard.

Of course we all would have much preferred to get the bull in first bite but it wasn't to be, as he'll be there shortly –keeping all our adrenalin glands stored, ready to explode when required. We handled, drafted and sorted that mob of 15 odd head, mostly quieter cattle as often occurs but then returned back to the receiving yard for our adversary.

Tony, (our for mentioned friend) very conveniently, was the last to move toward the receiving yard with the rest of us towards our ferocious friend, confirming he had the job to open the gate into the forcing yard but also, and more importantly, because he was now in front of the cattle, had to in fact come around to the rear of the mob – but on the outside of the receiving yard fence – how very convenient.

But as only a bushman would understand, there are certain unwritten rules in the bush. Everybody was very respectfully waiting for Tony to join us at our position, after all we didn't want him missing the fun as he cruised around outside of the yard very casually rolling a smoke. This was taking an astounding amount of extra time, and I just got the feeling Tony would prefer to be somewhere else at that moment in time. But we waited, and eventually he finally arrived to join us at the rear of the receiving yard, ready for our foray forward.

"Righto" somebody said, it was probably me as I was as keen as hell as all young bush kids are for excitement and challenge. We all started forward very gently with limited movement, as to NOT be the one to get the bulls attention, and enhance the chance for that individual, to be the object of his charging awareness.

The mob started to swirl again and stir up and we just hung back, even a meter or two towards the forcing yard would be a win in such circumstances. The scrub bull was constantly on the outside of the writhing mob of some 100 head of cattle, and although normally that would be enough cattle to hide a bad bull inside of –this bloke was having none of it. He was circling widely and faster than the rest and was going to have somebody, during his attempted escape back to the pastoral rangelands - that was his opinion anyway. He firstly came out on Johnnie's side, only three meters from me, but fortunately for me had Johnnie sized up for reasons only known to himself, and proceeded to drop his head and rush at Johnnie with a huge scary flurry, as the rest of the mob took his example and followed suit breaking back in the yard. Johnnie had completely stepped sideways and given way to the raging bull's demand – quite understandable really.

This happened a few more times, and by that time everybody had the adrenalin levels running on max settings, as every time he went by each of us, his fear was minimising and his cheekiness and anger increasing, and the distance between us and this freight train was closing, the big horns in very close proximity at times.

The old yards, surprisingly enough were not build for access through the gates by even a vehicle or "Bull Buggy" (a 4WD vehicle set up with smash bars surrounding it used for mustering wild cattle in the bush that often charge and smash into it). This vehicle would normally be introduced at this stage into the receiving yard, but the gates were too narrow. Not only to provide that extra different pressure for the bull and entire mob, but also to act as at least some form of protection for those gutsier individuals in the centre of the yards, and furthest from the fence for protection.

The more the unsuccessful attempts are made usually the worse the situation gets, especially as the animal gets what's called "hotter" and their reasoning capacity is minimised dramatically. They often start just charging directly into the yards at this stage, by hurling themselves into a jump, just prior contact with the yards and providing the yards are substantial enough, fall backwards on their backsides or side. So it was that he had done this once already with one of the boys, as he realised at the last moment that the animal was not only charging him, but continuing to make good his escape by crashing through the fence barrier the jackeroo had just fled to. To his credit the young man allowed nature's strongest drive of self preservation to take over, and quickly moved slightly further along the rail, only centimetres from the fast moving freight train. The shock of the huge weight smashing against the top rails had shaken the whole southern section of yard, and the boy knocked free from his grip was thrown down onto the ground, alongside where the raging bull had just landed, on his side and unbalanced with shock. The boy very quickly made his retreat for safer sections of fence, before the bull quickly got to his feet.

"Why didn't you throw the bastard while you had him down", somebody sarcastically yelled from a considerable safe distance – probably Tony. There's nothing quite like a bushman's Aussie humour whilst everything is hitting the fan. The young bloke sat perched and gasping for breath on the fence, like he had just sprinted 5 kilometres for his life.

I wonder how Occupational Health & Safety would have gone back in those days.

So it was getting serious now and we all knew it. But the job had to get done! The boss said, "Just sit back here on the back rail for a bit, and let things cool down". Nobody objected to that idea!!

So we sat for a few minutes to gather our senses and let the pressure off the cattle. The bull still patrolling the outside of the mob and throwing his head disgustingly in our direction, blowing from his nose and now dribbling large amounts of saliva, dropping to the ground in long, clear, stalactite like stems of liquid, now mixed with blood and mucous.

We sat there for some five to ten minutes I guess, considering our options, allowing the bull to go into the forcing yard himself, without any attention to us, to potentially explore another means of escape. As there were cows over in the other yard that way, and we hoped that may help our cause. The bull was exploring along the receiving yard fence, and then followed its length on into the forcing yard. Without a word required, six stockmen rushed from their crow like positions on the fence, towards the mob and the forcing yard gate. The cattle were all startled of course, (you normally wouldn't do that but this was an emergency) and more packed into the small forcing yard including our monster, making it too tight, and some had to be let back. Everybody was extremely conscious of the bull's position, now in the forcing yard finally where we wanted him, and even so he was initially in the running for an exit past us with the overflow cattle. There were three or four very keen stockman, who at last had a gate between themselves and this monster, and we were not going to let the lucky opportunity pass by without a fight. We held it with our entire mite, drafting the last few slipping stragglers back into the forcing yard pen, where the disgruntled bull was fortunately looking the other way for that mistaken instant.

The chain was put around the upright in quick time, I give you the tip, and a double half hitch was used for extra security.

"Wow", he was bloody huge, up so close in the forcing yard, now we could reach out and touch his huge frame, blowing saliva and blood over us as he swirled in the new, more confined environment pushing other animals into a swirl of the mob with his huge body.

He was constantly looking up at everybody as he rotated in the new small confines. A bull with his head always raised and looking at you and escape, is giving you a message and you don't have to be a brain surgeon to get it.

Despite the drama this bull had caused, we had all witnessed it many times before (perhaps not quite as big as this bloke) and our mind was on loading and getting the truck away as we moved forward of the forcing yard towards the raceway and crush to allow the bull and other cattle to settle down a bit, but then to enter forward up the raceway in time – and also to get out breath back, from the adrenalin rush we had all just experienced if the truth be known.

Now this is where "Tony" comes into his own.

Conveniently missing in action for most of the prior proceedings, Tony jumped smugly onto the forcing yard back gate full of confidence now that the huge bull is comparatively entrapped in a smaller forcing yard. I'm sure Tony would have been giving himself most of the credit for the job well done, but I noted he wasn't there when we needed him on that forcing yard gate before! Missing in action –I think they call it.

Cattle can be handled from the outside of the forcing yard mostly, without persons needing to enter it for their own, or animal's safety especially with wild cattle. The shape of the yard with its triangular small size will normally mean cattle will run past you, on the other side of the rail in an attempt to escape from your perceived pressure, and away towards the raceway entrance where you want them, without much effort or danger to man or beast. Even the wild cattle will often follow with the others up the race, as you pass on the outside, and so it is that everybody is inclined to relax once cattle are into the forcing yard – especially Tony.

Now everybody realises that sitting on the top of the back gate in the forcing yard is not the most productive position to take up, as far as the overall operation is concerned. However there is a slight need at times, for minimal pressure from the rear of the forcing yard to move cattle forward into the race way, so there was arguably a minimal reason for Tony to support his case for that hassle-free position. And anyway he could run the show from up there (in his mind at least) and not have to do much, which was right up his alley.

The mob has decreased quite considerably in size now in the forcing yard, but the bull was hanging back from entering the raceway and still swirling his mob and as yet he was untouched. Down to the last six or eight head and the forcing yard was only 30% full, which left lots of space and room for movement.

Probably three or four meters between the edge of the swirling mob, where the bull was, and Tony's position very comfortably sitting with his knees bent up under him, on the top rail with his eyes down rolling his rollie on the back gate.

Nobody was particularly looking, but sometimes events just happen in a cattle yard, that means everybody's attention is "glued to the spot", in an instant, when something serious happens. So it was on this particular day when the bull reached a position, during his circular swirling anger, to see Tony's outline, perched so relaxed and unmoved on the back gate, and this target very clearly became the focus point of his anger.

In an instant the animal made up his mind, and now there was even enough room for a decent run up. Close to the finality of rolling his rollie cigarette, Tony had taken his attention away from the angry beast for a moment surprisingly, believing the job was done.

The bull straightened his back, as they do, glued his attention on Tony and gave an angry flick of his tail, and shook his head violently as he took off in no uncertain direction. All the body language was scary to witness; nobody was left in any doubt about what his intentions were. Tony hadn't lifted his eyes to see the menacing beast approaching until it was airborne and only a couple of meters away and on direct route for Tony's bent legs, nearly two meters off the ground. In the split seconds from start to finish there was nothing anybody could have done –barely time to yell a warning.

With the affect of gravity on the huge animal by the time it reached Tony, its horns and rock hard head smashed into the gate, only millimetres below Tony's RM Williams boots, whose heels had hooked into the second top rail. Smashing the gate backwards to the end of its travel some 100 mm, its momentum only stopped rigidly by a heavy duty chain attached to the railway iron upright. The huge impact hurtling Tony and gate backwards at an amazing speed and force such was the animal's ferociousness, that the force sent Tony backwards and also upwards it seemed, and he looked like a circus clown or scarecrow flying uncontrollably backwards through the air. He would have to be dead, gee you couldn't survive that.

Tony seemed to hold time as he in slow motion, reacted so coolly to the onslaught like a circus clown, drifting in a gentle relaxed aerial summersault and landed casually on his feet some three meters away, bending slightly at the knees to cushion his fall into the yard beyond. Rollie still in hand, and still in the crouched position he stood upright, licked his rollie, and laid it flat against its rounded face, complete with a beaming cheeky smile, much to the relief of the applauding crew.

The unsuccessful beast roaring and glaring, blowing snot and blood, all over Tony through the bent damaged gate, in complete distain and disgust, only meters away. Dragging back deeply on his now lit rollie Tony smiled to his would be attacker through the gate barricade like a Spanish bullfighter.

Not only did Tony sidestep what surely was sudden death, but he responded in such a cool, matter of fact way, landing so casually, despite the pressure, rollie completely intact after a complete summersault, and then had the gall and cheek to smile about it.

Try doing that on a Sunday afternoon at the beach.

It was well worth the enduring applause and an event significant enough to record.

The only factual evidence of the event still available today, a huge bow in the top two inch pipe rails of the gate, where Tony had been smugly perched and sitting.

But for those of us who witnessed the event, it was well worthy of a bushman's knowledgeable smile, and a nod of the head in temperate respect.

All the rest of the day was just a boring formality after that, the bull and the balance of sale cattle were finally loaded without much further ado, and trucks headed south to market.

Tony had the last laugh again and it was the topic of conversation around the camp fire for months to come. There were many spontaneous outbursts of laughter from the boys in attempted re-enactments as they recalled the incident.

I guess it takes all sorts of characters that we have to share our working days with.

Do you reckon this bloke wants to say something?? Look at the stance
"A Mickey bull," like the one above (I thank Miss Attica Grey for her photo) who is a scrub
bull who's missed a few musters and the consequent castration that would normally apply.
They are quick and fast and used to constant fighting between themselves, very quick
on the forequarter to rip and hook violently with that head and horns. I have seen them
open up a horses belly like a sardine can, with one quick flick of the head, and the horse
had to be destroyed immediately. Let alone the damaging potential to a person.

LEFT: She's saying a really love you
RIGHT: Keros my brother horn tipping, with neighbour Jamie Dempster
easing the animal forward, watched by truckie at rear.
Horns need to be tipped (only needs a few centre metres from the tip like cutting your
fingernails) so during the trucking journey animals don't damage or bruise each other

LEFT: When my x-wife came up to the bush she broke in and rode this lovely little brumby colt she called Nugget above. His temperament for a bush horse was quite amazing. He was later shot by a very foolish roo shooter who didn't check to see his brand and thought he was a wild brumby.
RIGHT: A poll bull up the wooden railed drafting race at Pindillya Yards on Ballythunna Station. Frank Shaughnessy built these yards in 1950

Brian Lucas (with whip) and Alan Walters pushing up into the forcing yard on horses at Middle Camp yards.

LEFT: Mustering some escapee cattle back across the Ashburton
River with the helicopter near Nanutarra Road House.
RIGHT: The "Pencilor" Derick Freeman has a very important job to keep the figures right; calves
marked, sex, details and numbers of cattle trucked, whilst the younger staff do the handling job.

Wild Brumbies running on Ballythunna Station, they were exceptionally well
bred, from Perth Cup progeny and extremely sure footed, and strong.

LEFT: Only hours before my helicopter crash talking to the owner of Red Hill
Station, Eric Ainsworth (on the horse) and his aboriginal jackeroo.
RIGHT: The art of "calf scruffing" for marking one on the head, one on the flank.

LEFT: A few in the yard.
RIGHT: Cattle just let out of the yard at White Cliff's cattle yards on Curbur Station in a good season.

LEFT: "Do you want a piece of me?" says the Black Angus pole mickey.
RIGHT: Bogged to the eyeballs on Curbur. Lucky she's not loaded

LEFT: The receiving yard at Moolarie cattle yards, where timber was laid horizontally between uprights to keep it aligned with a cut top rail. It worked well because cattle never charged it.
RIGHT: Overhead drafting of sale cattle at Middle camp portable yards on the Wooramel River.

Chapter 5

The Station tip

Oh they're the most exciting place to go for a young bush kid, the old Station tip not the rubbish so much, but the old vehicles, trucks, motorbikes, old wooden carts with the horses long gone, and even old bottles with marbles in the top, that rattle when you shake the dust out of them. You would never guess what you were likely to find. It was so exciting and anything was a bonus.

Old tobacco tins, been sealed for years, that when you opened them with caution, would still hold the mellow smell of ready rubbed tobacco from years before. Their internal metal, shiny and proud after years of caring. Old dark brown and black bottles some long and square, others thin and round. Some crystal clear, like they have just been washed by mother nature's rain, shine and reflect like crystal jewels in the sunlight, others brown and stained that only harnessed more intrigue and wonder at what they had contained many years before.

The old vehicles so great to be able to jump in and steer, doing a complete steering lock each way, as they were sitting on blocks without any tyres to inhibit its progress, which were long gone. One could drive like a racing car driver making all the right noises, it sounded real good. Plenty of space in the back for imaginary passengers, were told to "sit down and be quiet" while you drove over imaginary hills and breakaways. No seatbelts then. Some passengers had to stand and hold on tight to avoid falling over from the rough driving surface, like I'd see Dad drive through the bush. Proper luggage of old boxes and bits and pieces could be piled on, but care was necessary for angry scorpions and centipedes that were surprised by the unexpected exposure to sunlight, and lack of protection. You always grabbed for grip near the top of the old boxes rather than the bottom, where unsuspecting nasties could attach to your fingers.

Skipping between the "clumps of stuff" with juvenile excitement, missing the rusty Ardmona fruit tin cans and odd broken glass, and inspecting some weathered leather strapping and some old saddle gear, cushioning horse tail hair worn and tired in its felt covering, but still rich, black and curly. Wild Hops plants standing strong, tall and green, with red flowers standing protected by years of rubbish and left over jobs of by gone times. Partly welded metal frames, welded with old carbide gas and wire, protecting coat of galvanising burnt off to expose to the elements ugly brown rust.

Horse hobble chains rattled as you exposed them from dusty soil, and rusted buckles outlived their leather counterpart which now was very withered, dry and twistered. Duncan's small metal wax matches boxes were everywhere, as I knew the waterproof wax matches would strike into flame anywhere or on anything. I had once a few years before carried a metal box in my jeans for months knowing I could start a "billy fire" anywhere if required.

The old wooden carts were great too, standing proudly on their huge round steel wheels, with strong metal spokes, some with two wheels and some with four, which were big enough to cart bales of wool behind the camels, many years before. Now abandoned at the dump for deterioration and kids pleasure and imagination. You just had to be very careful of the old timber bench seats, which had cracked and dried and splintered badly, slicing into the unsuspecting, overexcited bottom.

I'd had a lovely day travelling to the tip two mile from the house, on my little grey pony "Gypsy", who was now tied, minding his own business to a mulga tree in the shade, and shaking the odd flies off with the toss of his main in the air and frustrated flick of his tail. An odd snort through his nostrils blue away unsuspecting flies that were annoying him, been drawn by the moisture seeping from his soft, velvet nostrils.

I had had a scare during the trip out though, as at one stage I had leant forward in the saddle to stretch my back, just in time to see Gypsy step over a slithering black snake. Gypsy didn't seem to notice in his gate and certainly didn't change stride, and I remember the snake didn't seem angry either or threatened, as you may have expected. He didn't try to strike or bite at all, just get clear and out of the way. What scared me most I think was the thought of what may have happened. I could have been cantering Gypsy a little later and he may have collapsed on me and gone down, throwing me or of course even dying, if he had been bitten. If I had not happened to stretch my back and be looking over the horse's wither at the time, I may not have noticed the close shave at all. Nobody else seemed to care anyway as Buster, my black and white border collie dog following behind the horse just stepped around the snake as I looked back, too late to explain a warning to him. He's seen it all before and barely bothered to look back as he trotted on with me.

I always could pick my way to the tip via the "big gidgee tree thicket", where one particular Gidgee tree stood proud and tall above the rest, a main trunk that went straight up, and another on either side hung out like huge arms from a giant. We travelled along the cattle pad all the way from the corner mill trough. Past the black mulga "Park" as we called it, where the water used to lay for weeks in a low lying area after rains, and which was covered in beautiful smelling, pink wild flowers once the water receded.

I knew the holding or cow paddock well because that was where I had to muster the milking cow early each morning if she didn't come in for milking.

Yeah going to the tip, was nearly more exciting than going to Perth the big city, for a bush kid.

Chapter 6

Mrs. Jones in the classroom

It was mid morning on the Station and I was starting grade one in the new painted school room over in the office block. Mrs Jones had us all line up and march across from the side of the homestead some 100 meters across to where the school room was the second last room on the eastern end of the office block. It was next to Dad's office and then the "store" room on the western end where the Yamatjis used to line up for their weekly food supplies from Mum. I remembered the very big bags of dried apples and apricots, which I enjoyed to chew on, and flour that was stacked up in the store two or three bags high, with lots of tin food on the shelves. Braised steak and onions and camp pie were always in demand and tin fruit, peaches and pears were always a luxury and well sought after. Capstan and Log cabin tobacco were the variety available under the front bench, along with cigarette papers and matches.

But the school room was at the other end of the office block from the store room and that's where we were heading. Myself and Mrs Jones's other pupils were all in line marching towards the learning of books, pencil and paper. She asked us to make sure we didn't tread on the heels of the child in front of us as we marched, left right, left right.

I didn't think this school work was very important really, as it was much more relevant to know your direction in the bush and be able to catch, saddle, and ride a pony and not fall off, than this school stuff. Mrs Jones was telling Johnnie to catch up and be quite, so I figured I better pay attention. When we reached the shade just outside the classroom veranda Mrs Jones told us to "halt" and we came to a very roughly obtained stationary attention. Once instructed to come into the classroom through the freshly painted white door, I sat in the single desk and chair in the room, and the rest took their places at their respectful positions to face Mrs Jones.

"Good morning class" she said, with happy bright disposition as we all replied, and she enquired if we were all feeling well. Asked us all in turn what we had been doing over the Christmas holidays.

I was wondering if Gypsy, my pony was hungry after this morning's feed I had given him, and Mrs Jones asked me if I was paying attention, as she wished to show us what lovely crayons and art paper we had for the start of the new year. If we would like to take our colourful reading books out, Jack didn't have his books so Mrs Jones had to let him borrow hers. Billy said he wanted to go to the toilet, but Mrs Jones said he would have to wait until recess, which wasn't long.

That was my first experience having a class with Mrs Jones. She taught four children, all siblings in that class room and around the station run for twenty two years. As well as being a mother in the bush without any of the normal amenities, such as air-conditioning or twenty four hour power. The school sets came up on the weekly mail and went back on the return mail truck on Thursdays, to be marked by correspondence school teachers, and then returned. There wasn't any power to have the heater on in the classroom on cold days, and that sometimes made Billy whinge a bit.

There was never a complaint from Mrs Jones, she continued to place ALL her children's needs in front of her own and kept a constant happy, smiling disposition always helping, always guiding.

I was an honour to call Mrs Jones my Mum. And yes I was the only one in that classroom at any one time, but the effort she maintained to develop that imaginary world she constructed to keep me even remotely interested, went way beyond the role of a remote teacher.

Mum in her younger day. Apparently she had twelve marriage proposals before dad became lucky, as she sung professionally to the troops in World War II.

My Mum, June Keynes, (Mrs Jones to her students) on the right with her mother in law "Nanna" Keynes.

Mum as a grandmother in later years holding Clare my middle
child when Alzheimer's disease took its toll.

Mealtimes were special, especially the evening meal when everybody washed
up and dressed appropriately and discussed the day's events around the table.
It was a time for family to relax and communicate – no TV

The front entrance to the Curbur Homestead, where I was born and spent 25 years.

The Curbur Homestead from the air looking South West, Mrs Jones's school room is in the bottom right hand corner. The shade house in the centre to collect breeze

Ready for a lovely meal at Simone's, my sister in law In Geraldton WA.

Who said you don't celebrate Christmas in the bush at Ballythunna, in the "bow shed" where water trickles down the grass walls to cool the inside once the breeze blows through. If started early in the day it can be ten or fifteen degrees cooler inside.

Chapter 7

The "Cool Kid" on the skate board

I would have never guessed that when I took the kids to the Glendenning Road Skate Park, in Geraldton, that the two girls would have the most exciting experience of their life.

I said to their Mum Krystal, I would take the three children out for the day, to give her a chance to catch up on some assignments she needed to complete before next week.

It was a lovely Saturday morning around ten am when I called for them and we proceeded to pack the picnic hamper of fruit, drinks and chips. I packed Robert's Animator bike in as he was four, Jasmine's Push tricycle and Ayla's Scooter for the two seven year old girls for the skate park. And I remembered thinking later "hell", I didn't bring their helmets.

We dropped in to Macca's on the way to the park for an ice-cream cone, but their machine wasn't working, so we backtracked to "Hungry Jocks", as I like to call them. They ate them with amazing hunger and the unusual quietness from the rear seat, confirmed they were hitting the spot. After a few wrong turns we ended up taking Glendenning road off Brand Highway and ran just around the corner to the Park. As we were in close proximity, Robert was shouting from the back seat, "that's it, that's the park, there it is".

It was a beautiful grassed park with a skate park on the southern end, with kids play ground area toward the north western end and a nice family BBQ area on the western side with big shade trees, where a large family were having and early lunch and kicking the football.

A little girl was trying was trying to kick the football back to her father but missed her foot completely as she through it down.

I pulled the car up on the eastern side of the park, in a gravel car park area with a large heavy duty chain spread across the entrance to stop unwanted access for vehicles onto the grass oval, but with a lock on it for the gardeners' tractor to service the park. We unpacked the boot of bikes and excited children rushed toward the skate park like a herd of stampeding elephants, dragging and pushing their bikes ahead of themselves as if the park was going to leave by the time the last child arrived.

I carried the picnic basket along way back at the rear of the excited children, as they skeltered off attacking the surrounds of the skate park with serious abandon. The skate park itself ran north south with the northern end containing the "bowl" around the top and "rails" on the eastern and western sides. The "middle pipe" looked like a metal bench between the two ramps with the down ramp on the northern end.

The girls were push running with the trike and scooter initially, but tired of that, as running away from Robert soon lost its novelty, especially as he was very proficient on his bike. Chasing them, then catching and flying past the two girls who quickly selected particular areas like the southern ramp and "Middle pipe" to be used as "barleys areas" so they could flee to quick protection.

And then it happened, not straight away, it took a little time, but something absolutely came into being.

A car pulled up alongside ours in the gravel car park and a mother was obviously bringing her son, a young boy, out to practice with his skate board on the park.

He casually walked over with his board and dropped it nonchalantly onto the concrete, jumped aboard and pushed with one leg up the small "Bowl" and around behind the drop. He pushed his board indifferently towards the drop, held it momentarily with the back axle catching on the ridge, and then placing both feet on the board, dropped it over the ledge and rode it to the bottom and beyond increasing in speed and poise.

He just oozed self-confidence and talent, and the two girls who were initially playing with Robert, had lost interest in him and came to sit with me on the park seat, giggling and laughing at each other, but not removing their eyes from their new young hero. They were squashed up together on the bench seat, and in turn on me, still laughing and giggling uncontrollably as if they had just been given laughing gas. I felt like there was something in the air. There was a magical see through mist of love and instant attachment to this handsome young stranger, with his long, blonde curly hair, and although not overconfident, was very casual and self-assured with lots of skill to boot. Well the girls certainly thought so anyway.

Ayla the older of the two girls said in a giggly girlie way, "he's very handsome Poppy", and Jasmine agreed wholeheartedly. It was beautiful to see the confidence the young man of about fourteen had with his board, he handled it like a third hand, flicking with his feet on its rear tip for it to rear into the air for him to casually catch it in a relaxed, very matter of fact way, ready for the next onslaught down a ramp or along a pipe. It was as if this was all part of the tools of his trade, like a top "grano" worker would be with his trowel.

The girls said they would love to kiss him and marry him and be with him forever, and that he was so very, very handsome. They crooned as they watched his every move, suffocating them with his presence from thirty metres away. The masculine fit young body and board in uninterrupted motion together as a complete unit, just like another body appendage, fast to the bottom, slowing to nothing at the top, only to drop away again, gathering increasing speed with the use of gravity.

And so it was that day I was so lucky to be able to witness one of the natural wonders of the world, and may it never die. It did an old man's heart the world of good, to be witness to such a beautiful experience.

The only downside was the trip home with two lovesick girls, who had their first taste of sweet magic, but Robert just said he thought they had "gone mad".

The Keynes cousins when younger.

The Keynes country girls older from left Clare, Krystal (sisters) and Amelia and Matilda (sisters) at their Cousin Katherine's wedding.

Michael and Matilda as youngsters learning to "boil the billy"
with pup in hand on Curbur at Chivas Hill.

Camping with Michael my youngest with his pet dog "Sammy".

LEFT: Annie Bates helping us at cattle muster with her horse at Middle Camp.
RIGHT: A young bloodstained Curbur jackeroo happy and helpful.

Kim my X wife of twenty years with her lovely blue healer "Blue" in the wild flowers.

A chance for the crew to have a dip at a river pool near Middle Camp on the Wooramel River.

Kim up the top of Mount Augustus look out in the upper Gascoyne.
She got up there the easy way in the chopper with me.

Krystal on the bike with me on back lawn at Ballythunna Station, where she was born.

Chapter 8

What a lovely gesture to do for a stranger

It was late afternoon and I had been around town dropping my CV into any perspective employers and contacts I had, and realized I was dry and the last 10 dollars I had to my name, was in my wallet.

No credit cards or term deposits –that was it!

It had been ages since I had had a beer because of my budget constraints and I decided to go into the GBH (Geraldton Beach Hotel) to be able to buy just two middies of my favourite swan draught on tap for the trip home. I knew it was a bit foolish, with nothing in my fridge at home or barely enough milk and bread for a day or so, but I just felt like "bugger it", what will be, will be – I felt it was a desperate time for a bit of positivity to intervene into my life!

I had been out of work for nearly six months, an offshoot of the casual employment in the Mining Industry and my meagre budget had been blown out of the water weeks before, with savings gone and no complete knowledge of what was to be.

Now I'm not a great "Boozer" actually, not now days anyway, perhaps when I was younger and working in the bush we'd knock over a few as you do, but I certainly couldn't call myself a regular. You know when you just get worn and tired of fighting the inevitable, trying to constantly swim upstream, week after week and things just don't seem to be falling into place. In fact quite the reverse it seems.

So this was the mood I was in, as I approached the front bar for my first, swallowing the thick dryness in my throat. As I waited for the "skimpy" to serve me, she was bright and happy with the crowd and little did I know that serendipity was to follow her lovely smile that warm afternoon.

The first mouthful I swallowed was cool and satisfying and I drank eagerly, and I tried to enjoy it to avoid me calculating the awful fact that half my budget was gone. I pushed that thought to the back of my mind and continued to enjoy the light amber liquid as it quenched my thirst and tickled my enjoyment. And it was time I got out too. Hadn't been anywhere for ages, justifying the nagging whinge of budget constraints.

It wasn't long before Greg Barnes came in alongside me and gave me a tap on the back in saying hullo. He was a big happy man who was mostly smiling and was well known for helping young skimpy's out when they were in a bind. He was a regular and well known and appreciated by the Management (mostly female) Just recently he had driven a girl to Perth, some four hours drive when she missed the bus and needed to attend some very urgent appointment. For memory she may have even been a tourist out in the Australian regions, extracting another year on her visa for working in the bush. From Greg's viewpoint he was just helping somebody out who needed a hand "Aussie style".

I was torn back into reality by my empty glass, and the skimpy throwing coins in her jug, and moved forward to place it on the bar for a refill, apologising to Greg I couldn't buy him one as I had occasionally done in the past. We sat down alongside each other on bar stools and developed a conversation of reasonable importance - as men do in such cases, and was enjoying my second and last, just starting to feel more relaxed now and being present with the moment and atmosphere. A fair crowd was in now and the noise level had gone up a few notches, and I had to lean closer to Greg to catch his conversation as a few other mates of his came in, and made contact with him in their usual ways.

I was leaning my head back to catch the last drops of my second middy, as Greg was joking with the manageress and ordering a round for him and his mates. He said, "Do ya want one?" in my direction as the barmaid placed his order on the bar, and I blurted something feeble out like I had to go, or couldn't return the favour. "Get 'im one", he smiled to the barmaid, as he nodded to me, and he passed the frothy beers back to the other boys. "Don't worry about it mate" he said, it happens to all of us, "we've all been there at some stage". The knowingness in his smile said it all; you knew he was talking from considerable experience, and yet his understanding and kindness was very powerful.

I think that was the nicest beer I had enjoyed in a long time, a beautiful, simple but sweet Aussie gesture, I had not experienced for quite a while.

Nobody cares when you're at the bottom of the barrel do they?

As I sat there enjoying that gift I realised just how important it was to give, and coupled with the big knowing smile on Greg's face he gave me, made me actually appreciate how lovely it was to give, as it was to receive. A demonstrated lesson I will always remember.

My inhibitions melted away and I think Greg bought another before I left, realising I was trying to maintain some sort of responsibility for the driving trip home. Somebody had just told a joke and we were all enjoying the punch line, when Greg very inadvertently and without warning reached across and placed what seemed like a note in my top pocket. He stopped me as I went to reach for it and said, as half expecting I may take it out to give back to him, "Just leave it there, I figured you may need something" . I was actually quite embarrassed, but the few beers had hit the spot by then, and instead, the sweet little pangs of niceness, comfort and gratification took over and filled the pit of my stomach.

I said good night to the boys and walked to my car in humbled gratification, and the overwhelming warmth of giving controlled my body, filled my senses and quite overcame me. Months after the experience I still become quite emotional from the episode, in the nicest possible way. I guess that's just what giving does, quite overcomes you.

Greg was a taxi driver, and I knew through previous discussions with him it didn't earn him huge money, but he was still happy.

That night I was preparing a meagre meal in my little kitchen and without reason inadvertently, reached into my top pocket for an unexplained scratch. My concentration from the meal was diverted, as I realised it felt like money in my pocket, an unfamiliar experience in recent times, it was only then that I looked down and proceeded to withdraw not one, but two fifties from my top pocket. Tears flowed into my eyes and for the second time that evening I was quite overcome by emotion. What more can one say, there is amazing human beings around, and what a great time to run into them when you are on the bones of your bum.

So if you happen to get a taxi in Geraldton WA, and the driver is a big happy fellow by the name of Greg, give him a big tip anyway, and tell him the next beer is on you.

Some of the lovely sights of the Murchison Pastoral country in WA.

LEFT: A lovely little pool in the Tin Hut Creek which runs into the Curbur Lake.
TOP RIGHT: Belung Pool, pronounced "belang" just north of Byro.
BOTTOM RIGHT: The two mile breakaway on Curbur, containing some caves
where kangaroos camp in the cool but with the best view from the high country for
protection from predators. The type of site where dingoes have their pups.

Looking south over the North West corner of the Brille Brille Lake on Curbur
Station. We used to water ski here for months after it rained, many square
kilometres of open fresh water. This is on private property don't forget.

LEFT: A typical Pilbara "rock hole" against the rich red rock of the Pilbara.
RIGHT: See my co-pilot on the LHS??? Pure bred Border collie "Ben". I could talk to
him on the ground via loudspeaker when I was mustering sheep or goats. He could
muster a blowfly into a bottle every time because he was such a good "eye dog".

LEFT: Fork lightening flashing outside the Ballythunna Homestead during a thunderstorm in 1990.
RIGHT: Another shot of Belung Pool - permanent fresh water
which has never gone dry, even in the worst droughts.

LEFT: Bull camel in a good season with Mount Augustus or Burringurrah
in the background, the largest rock in the world.
RIGHT: Another picture of the Brille Brille Lake on Curbur, many square
kilometres of fresh water when it fills to 5 metres deep.

The following are images of when my 2 seater R22 helicopter flies.

The bottom right image is when it was parked on the front lawn of Brickhouse Station, Carnarvon.

The following are some images from my near-fatal helicopter crash near Red Hill Station in the Pilbara, W.A. on September 29, 1984.

This is me lying for 16 hours on Lionel Ainsworth's lap, with my pilot Rodney Pulford standing at back left, who saw me go down and came over and pulled me out of what was left of the machine before the RFDS (Royal Flying Doctor Service) came in with a chopper and doctor to stretcher me out. I was approximately 150 feet from the bottom of this gorge when my engine failed above the very rough terrain.

Placing me into the rescue 206 chopper in stretcher.

Some 10 years later after this accident I organised the annual Ballythunna Bush Bash(BBB) on my Station property to raise thousands of dollars for the RFDS where 700 patrons came from all over WA to attend. The highlight of the weekend event was the 5K prize for the Brumby Colt muster, where the brumbies were widely admired for their speed, type and temperament.

Chapter 9

The annual Ballythunna Bush Bash and Brumby Colt Muster

This was a real pleasure to organise and be involved in, giving something back to the RFDS for helping me the way they did to have such a unique event in the real bush where people of all walks of life could camp out in a picturesque environment with very good temperament cattle and horses. Where only last night (8th March 2014) at a BBQ a young man mentioned to me how his father had attended the last BBB event in 1995, some 19 years before and was still raving about what a great experience was had by all. What a great honour to have for me, all coming from me been involved in that awful helicopter accident in Sept 1984, where the RFDS (Royal Flying Doctor Service) and their great committed staff and crews had removed me in a stretcher from a gorge near Mount Farquhar in the Hamersley Ranges and flew me to medical attention in Perth at Royal Perth Hospital where I required spinal fusion.

But it makes you realise how lucky you are. After specialists not knowing if I would walk or have children, 30 years after I'm very fortunate to have three beautiful healthy children and 2 grandchildren (pictured above) and at 58 years young I'm just starting to ease back on the tail of the main mob – and perhaps write an odd bush story or two!!

But I am pretty driven about Agriculture and its future, and where it should be on the world stage.

I hope you may be able to take and enjoy some of that pleasure demonstrated by these pictures of the BBB (Ballythunna Bush Bash) below.

Note the delight and involvement from the fellow competitors and helpers, who were not professional rodeo riders, but just wanted to have a good time.

Before this weekend below took place in 1995, the location of this unique site nestled into a natural amphitheatre in the Ballythunna Range was completely isolated and hardly seen by white man.

During the few weeks prior we set up complete male /female ablutions including flushing toilets and hot water, live band and music, huge big top tent, 240 volt power, a large range of cold drinks from the bar, public campfires and a damn good weekend, where lawyers sat with stockmen and kids played in the hills . We did all this completely on our own budget and no government or private funding. The local community chipped in.

This was the ad that started it all off in 1993.

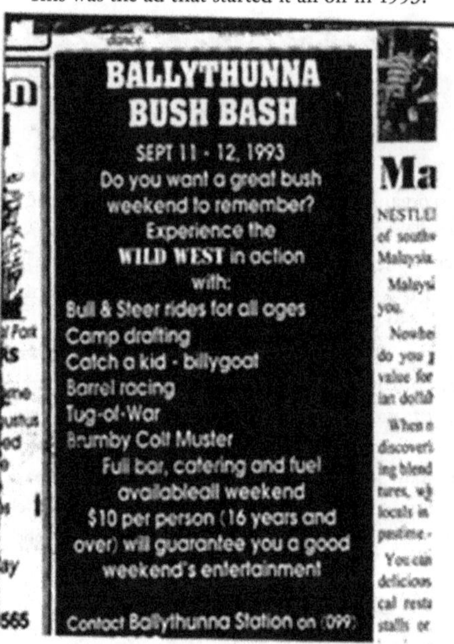

Just in case of emergency the RFDS were kindly in attendance all weekend, with medical staff and above aircraft on standby on the Ballythunna Strip.

City kids were given enough confidence and support to have a go at the calf riding (after the mutton busters) with lots of encouragement. I used to be able to hold the calf's head until they were confident and the quiet temperament of the cattle that I bred made it safer and more enjoyable for all. There was much emphasis placed on getting children involved into these country activities. I think one of the secrets to why they still talk about the BBB 20 years later.

I was riding my motorbike up the river one day on a mill run and thinking about what events we could develop for the BBB. I rode through the river and saw a heap of "paddy melons" as large as watermelons. They worked well for relay races and throwing of the small ones for the youngsters above into a bin under the clock was a great distraction.

The "calf scruffing" in teams of two, had to throw the 100 plus kilo calf on its side (one on the head the other on the flank) for marking but on this occasion apply a paper tail tag on the animals' tail in the fastest time, which was when I dropped my hand for the timekeeper.

Going – Going – GONE!!!

"Spin" talking to a lady, but Gary's got him covered.

Gary Hancock kindly travelled to the BBB to demonstrate "Spin on Command" for us, his cutting stallion. Also because the cattle were very quiet they would work right up to "Spins" nose nearly. I got somebody to put video of Spin cutting these cattle to classical music – oh it's priceless.

Gee whiz, he's handsome – oh, sorry, that's me!

There's the "Brumby colt", first time in his life that a human hand has touched him. He has just galloped 30 plus kilometres with his mob after we started him and I selected him as the nominated brumby colt for the 5K prize money. This is also the first time in three years that the horseman even got close to the brumby colt. Lots of credence goes to my brother who blocked him up with the bull buggy prior with the channel 7 camera crew.

The chopper that was used to locate the brumby colt above and
run him and his mob into the waiting horseman.

The BBB included some dog lovers, a few hot young night dancers,
and some young riders learning the ropes.

People that are not aware of the natural grazing of the rangelands often wonder what livestock eat, especially when feed is not that obvious. In the above photos if you note the foliage and leaf matter on the top feed or perennials (the trees and bush that can obtain up to 12% protein or more, they are the haystack) and the "annuals" are a bonus when/if it rains. The above few photos I could find are not the best, but at least gives you a better grasp perhaps of what a reasonable season actually looks like. It does the heart good to see it and know your livestock have a good season ahead and some food security, as well as knowing that your females will increase body weight and there will be lambs and calves on the ground everywhere in the near future.

These are some lovely wildflowers(annuals) with purple vetch (which has a beautiful scent) and white everlastings. The bush usually comes alive with birds, insects and the associated noises when it looks like this.

This is the lovely rose garden we grew at the front of the Ballythunna homestead in the late 80's with scented roses.

In conclusion, here are some shots of the lovely Ashburton River, where I was working with an old neighbour mate Joe Armstrong who is managing Uralla Station for BHP.

The Ashburton River in flood and rising at the Uralla crossing, not far from its exit to the Indian Ocean. Don't you feel like you're there??

The truck that didn't make it. A cement truck attempted to cross the Ashburton,
but slipped off the concrete crossing on the passenger side – no rails fortunately
it didn't roll, however it needed to be pulled out by a grader.

Something completely different: The Murchison ASKAP (Australian Square Kilometre Array Pathfinder Radio Telescope) project on Boolardy Station. I worked for a subcontractor and we built 3 or 4 of these twelve meter towers, but unfortunately the paint applied in China was of inferior quality on reflector surfaces, so the contract concluded much earlier than originally planned, and I moved onto Mining work. The above project was opened Oct 2012.

From the most high flying technological project in the Murchison
above - to the most basic natural picture below.

The Murchison Shire Council (well known as the Shire without a town) has a population
of 113 people, 29 Station properties and an area of approx 50,000 square kilometres.
Please look at the photo above – and take the time to immerse yourself in it. If you have suitable
technology you may be able to go down and quietly imagine sitting under one of those millions
of trees in these beautiful pastoral surroundings and just meditate there for a while on all things
natural. You will be the only human being within 50 kilometres!! This is the big difference between
the "real bush" and other Agriculture, or even other lifestyles –it's just you and the real bush. Some
people understandably find this isolation frightening, but others enjoy its freedom. From a very
young age it makes you realize that you are just a drop in the bucket in nature's big picture.
And finally, I am so grateful that you have taken the trouble to read and hopefully enjoy the above
stories and pictures, (the first of many more I hope) not so much that it involves our family, but just
that we happened to be the custodians at that time, giving you a better understanding of the outback.
I wish you well.